BLACK BAT FLOWER

ASHLEY GISH

lok!

BLACK BAT FLOWER

CREATIVE EDUCATION • CREATIVE PAPERBACKS

Published by Creative Education
and Creative Paperbacks
P.O. Box 227, Mankato, Minnesota 56002
Creative Education and Creative Paperbacks are imprints of
The Creative Company
www.thecreativecompany.us

Design and art direction by Blue Design
Edited by Kremena Spengler

Photographs by Images by Alamy Stock Photo/Di Mac , 5, Di Mac / Stockimo , 17, electra kay-smith , 21, Itsik Marom, 22, mustbeyou , 20, Nadine Menezes, 2; Getty Images/ Photo by James Keith, 9; iStock/12; Shutterstock/Bignai, cover, 15; Wikimedia Commons/ Cameroncomix, 1, geoff mckay, 6–7, Geoff McKay, 18–19, Jdasiii, 11, Krzysztof Ziarnek, Kenraiz, 14, Leyo, 14–15, Nathasha1996, 13, Ton Rulkens, 16, 23
Every effort has been made to contact copyright holders for material reproduced in this book. Any omissions will be rectified in subsequent printings if notice is given to the publisher.

Copyright © 2026 Creative Education, Creative Paperbacks
International copyright reserved in all countries.
No part of this book may be reproduced in any form without written permission from the publisher.

Library of Congress Cataloging-in-Publication Data
Names: Gish, Ashley, author.
Title: Black bat flower / Ashley Gish.
Description: Mankato, Minnesota : Creative Education and Creative Paperbacks, [2026] | Series: Look! | Includes bibliographical references and index. | Audience: Ages 6–9 | Audience: Grades 2–3 | Summary: "An exotic plant with wings like a bat and whiskers like a cat? Elementary-level readers will learn all about the tropical black bat flower with this striking introduction to the unusual plant's life cycle"– Provided by publisher.
Identifiers: LCCN 2024044291 (print) | LCCN 2024044292 (ebook) | ISBN 9798889895848 (library binding) | ISBN 9781682777503 (paperback) | ISBN 9798889896647 (ebook)
Subjects: LCSH: Tacca chantrieri–Juvenile literature. | Tacca Chantrieri–Life cycles–Juvenile literature.
Classification: LCC QK495.D54 G57 2026 (print) | LCC QK495.D54 (ebook) | DDC 582.13–dc23/eng/20250115
LC record available at https://lccn.loc.gov/2024044291
LC ebook record available at https://lccn.loc.gov/2024044292

Printed in India

TABLE OF CONTENTS

Black Bat Flower 8
Growing Bat Plants 11
In the Ground 12
Leaves and Flowers 14
Spooky Flowers 16
Why Whiskers? 18
The Cycle Continues 21
Nature Stories:
Many Uses 23
Websites 24
Read More 24
Index 24

BLACK BAT FLOWER

The black bat flower is a strange-looking plant. It's named for its black **bracts**. They look like bat wings. It's also called devil flower and cat whiskers.

Bat flowers grow in Southeast Asia, India, and China. They grow all year long. They live in damp, dark parts of the rainforest.

bract: leaf-like petal under a flower

There are at least 15 kinds of bat flower, including white and green.

(10) Rhizomes grow into full-sized plants faster than seeds.

GROWING BAT PLANTS

Seeds

Bat flower plants grow from seeds and **rhizomes** (RY-zomes). Berries grow in pods after the flower blooms. The berries have seeds inside. After the pods dry out, the seeds are ready to plant.

Rhizomes grow from a parent plant under the ground. These copies of the parent plant can be cut and grown apart from the parent.

rhizome: underground stem from which new plants can grow

IN THE GROUND

Inside the black bat flower seed is an **embryo**. There is a hard shell around the embryo. This is the seed coat. Warm water softens it.

The seed must be kept warm and moist. Up to nine months later, the embryo breaks through the seed coat. Roots grow down. The stem breaks through the ground.

embryo: part of a seed that grows into a plant

LEAVES AND FLOWERS

14

Plants grown from seeds are called seedlings. Seedlings grow up quickly. Oval-shaped leaves grow on the plant's **stalk**. Black bat flower plants grow to be more than 3 feet (1 meter) tall.

stalk: main stem of a plant

After two to three years, flowers grow from the tip of the tall stalk. Each plant can make up to 25 flowers. They bloom from February to October.

SPOOKY FLOWERS

The most striking part of the plant is the flower. Each flower has three **sepals** and three petals. They curl backwards from the middle of the flower.

The bat wing-like bracts are surrounded by up to 26 long, greenish whiskers. The whiskers are small bracts. Fruits grow in the middle of each flower.

sepal: the green outer covering of a flower bud

WHY WHISKERS?

Scientists don't know exactly why bat flowers have whiskers. Some scientists believe the whiskers attract **pollinators**. Pollination helps the plant make fruit and seeds.

Blood-eating insects called midges may see the whiskers and think the plant is an animal. The midges look for blood by crawling past the whiskers. This may help spread pollen.

pollinator: insect that spreads a yellow powder, called pollen, from one flower to another

The black bat flower stays leafy and green all year round in the rainforest.

20

THE CYCLE CONTINUES

Each black bat flower starts as a seedling or rhizome. Roots grow down. The stalk grows up. In a few years, the plant grows flowers. After five to seven weeks, they dry up. When the seed pod falls to the ground, a seed may sprout. Grown plants send out rhizomes. The cycle continues for many years.

Life Stages
Seed/Rhizome
Root
Seedling
Stalk
Flower
Fruit

NATURE STORIES: MANY USES

All parts of the black bat flower can be eaten. The leaves have a bitter flavor. In some parts of Asia, the leaves are cooked in a tangy sauce. Medicine can be made from the roots. Some scientists have found that medicine made from the roots is good for heart health.

WEBSITES

Kidz Herald: Discovering the Mysterious Black Bat Flower
https://www.kidzherald.com/discovering-the-mysterious-black-bat-flower/
Learn more about where the black bat flower lives and how to grow it.

Soft Schools: Black Bat Flower Facts
https://www.softschools.com/facts/plants/black_bat_flower_facts/934/
Find out fascinating facts about the black bat flower.

READ MORE

Einstein, Tamara. *Weird Plants*. Tukwila, WA: KidsWorld Books, 2021.

Kaiser, Brianna. *Weird Plants*. Minneapolis: Lerner Publishing, 2023.

Lundgren, Julie K. *Creepy but Cool Scary Plants*. New York: Crabtree Seedlings, 2021.

INDEX

bract, 8, 16
embryo, 12
medicine, 23
midges, 18
pollinator, 18
rainforest, 8, 20

rhizome, 11, 21
seed, 10, 11, 12, 14, 18, 21
seedling, 14, 21
sepal, 16
stalk, 14, 15, 21